TOM WALKLATE

Greatest Racers

Stories of the Greatest Ever F1 World Champions

Contents

1

Introduction:

Welcome to Greatest Racers! If you are as much of a fan of racing as I am then I have no doubts that you will enjoy reading this book.

I have been watching Formula One since I was young and grew up watching some of the names in this book. Some of them were before my time but I still have great admiration for how the past greats shaped the sport into what it is today.

This book is a credit to the all time greats of Formula One, or F1 for short. The book includes the top 10 drivers post World War II where F1 became a recognised formula in the 1950s. The list is based on win percentages, their number of world championships and their dominance during them, as well as their influence on and off the track.

The list you are about to see is organised in chronological order and is by no means a ranking from best to worst. I also appreciate that this book is not an exhaustive list and that some people may believe that there are drivers missing that should be included. However, based on statistics and a trickle of my own personal opinion, I narrowed it down to 10.

I hope you enjoy reading about the greats as much as I enjoyed writing about them.

2

The Early Days

F1 was recognised as an official formula in 1950. World championships only had a few races and official rules on car specifications were still being established. In the first years of F1 it was regarded as the most dangerous sport in the world. With extremely unsafe cars with very little in the form of safety or protection for the driver or the spectators for that matter. Drivers were almost expected to die during a season. This meant drivers raced with an extra risk and danger hanging over their heads, making driving on the limit to dominate the competition even more impressive.

Motor
For the
Best Reports
and Pictures

3

Juan Manuel Fangio

Argentinian born Juan Manuel Fangio came from humble beginnings. Fangio came from a small town in Argentina called Balcarce and was son to immigrants from the Abruzzo region of Italy, a place in which he was very fond of. His integrity and discipline was credited to his hard working parents.

A mechanic by trade, he spent 40 years in the industry, beginning his racing career driving poorly self-prepared cars in long distance races across South America. His later promotion to F1 would seem much easier compared to the challenges he faced in these particular races.

Fangio was the first to dominate the world championship and has been considered the greatest of all time for a long time. Winning his first championship in 1951 with Alfa Romeo he was destined to be a significant force. His exciting driving style was a spectacle for the trackside onlooker as he was known to throw the car into the corner and wrestle it in a four wheeled powerslide and emerge out the other end in a cloud of tire smoke. Possessing a huge strength he was able to maintain this ferocious driving style all race which is a feat in itself, especially with the races back then which were endurance races lasting over three hours.

Of the 51 races he was in, 48 of those he started in pole position and had a win percentage just under 50%, the highest ever. Of the five titles he won, four of these were with different teams using different cars. Known as the 'Old Man' Fangio won his last world championship in 1957 aged 46, with most of the competition being much younger than him as well as coming from much more affluent backgrounds, it is hard to believe Fangio was as successful as he was

4

Sir Jack Brabham

B orn in a town just outside of Sydney, Australia in 1926. From a young age Jack Brabham developed a keen interest in motor vehicles by learning to drive and fix his fathers greengrocer delivery trucks. After failing to remain in college studying engineering he began working in an engineering shop and then a garage. At 18 he joined the Royal Australian air force in Adelaide where instead of learning to fly like he wanted, he was used in the mechanics division who were short of men with know how. When he was discharged in 1946 he opened his own engineering garage.

In 1948, Jack was introduced to racing in a series called midget racing which is very popular in Australia and involves racing powerful, lightweight cars around an oval dirt track. It was very dangerous given the design of the cars, with many of them being a shell and an engine with very little protection for the driver if they were to crash. Jack helped build his midget racer and became a regular winner, winning 4 successive championships. With a burning desire for more he came to England where he met Charlie and John Cooper who were the designers for the Cooper Car Company and designers of his formula one car that started his career. The design, inspired by Brabham, was rear engined, opposing the preferred front engined racers, this step was to change the sport indefinitely.

In 1959, Jack got his first F1 win in Monaco in his Cooper on the first race of the season which led to his first championship in 1959. The championship came down to the last race of the season at the US grand prix with rivals Stirling Moss and Tony Brooks all in contention for the title. Moss retired and Brooks finished third with Brabham pushing his car over the line to finish fourth when he ran out of fuel, but clinched the title by four points. In 1960 Brabham was victim to a crash in the first race but revived his season with five straight victories to win his second title in a row.

In the seasons following his back to back world championships Brabham's Cooper began to fall behind the opposition with the likes of Ferrari, Porsche and Lotus cottoning on to the transition from front engined to mid engined racers. Brabham left Cooper in 1962 and began driving for his own, newly formed team Motor Racing Developments or MRD for short. The MRD Brabham cars competed in the lower division of F2 and were very successful and were slowly catching the pack in F1. In 1966 came a change to the F1 regulations introducing 3-litre engines which suited Brabham much better. After persuading an Australian automotive components company Repco to produce an F1 engine for his car, Brabham returned to F1 aged 39 to win the world championship for a third time in a car he designed himself. This is a feat that has not been repeated to this day.

Jack Brabham's final grand prix win came in 1970 in South Africa aged 44 where he retired at the end of the season. The successful Brabham team was sold and he retired to Australia where he ran businesses in farming, aviation and car dealing. In 1985 Jack was honoured by the Queen of England and became Sir Jack Brabham due to his contributions to motor racing. Jack Brabham died at home in 2014 aged 88.

5

Jim Clark

Born into a family of farmers in Scotland in 1936, Jim Clark was raised in the Berwickshire hills on the border between England and Scotland. Racing was not in his blood with his family swearing to use them for farming purposes only. When he was a teen Jim used to take the family car out around the fields in secret before he passed his driving test and bought a car in which he entered into local rallies and skill tests. With the encouragement of wealthy friends he began to win a variety of races and competitions in a large range of cars. However, Jim Clark did not like the limelight one bit in fact he felt embarrassed by it even when people started to gain more and more interest in his career.

After being invited to race with Lotus at junior level he soon impressed Lotus owner Colin Chapman and was promoted to the F1 team within two years. His first race was Spa in Belgium nearing the end of the season which was one of the toughest weekends in history. Chris Bristow crashed fatally in front of Jim causing Jim to have to swerve around Chris' mutilated body splashing his Lotus in blood. Later in the race Jim's teammate Alan Stacey was also killed, this race weekend almost made Jim retire.

During his first full season another fatal crash almost made Jim retire caused by his collision with Wolfgang von Trips in the Ferrari which also killed 14 spectators. However, Lotus designer Chapman convinced him to stay. This duo would then become one of the most successful designer driver pairings in the sport. Jim would rarely be beaten on the track only really due to mechanical failures as the Lotus was unreliable. These mechanical failures meant that Jim lost the 1962 championship in the last race due to an oil leak and again in 1964 because of the same problem. But when the reliability was good Jim would walk the championship, winning the majority of the races.

Jim Clark's fame would reach across to America when he won the Indianapolis 500 but Jim was still shy to the limelight, avoiding press conferences and being visibly uncomfortable when speaking publicly. Although, his peers on the racetrack stated that Jim was the opposite and in the car he was calm and had an unmatched controlled aggression that made him a force to be reckoned with.

He never had ambitions of being one of the greatest and he always dreamed of settling down and having a family in Scotland, purposely only signing one year contracts so that he could leave whenever he wanted to. Tragically though, this dream would never come around because in April 1968 Jim Clark was killed in an F2 race in Germany when his Lotus had a tyre failure. This was a shock to the sport of racing as he never made mistakes and people felt that after his death the sport had lost its heart.

6

The Glory Days

With safety regulations improved the high death toll associated with motor racing and F1 was slowly on the decline. With the races now televised around the world, F1 was becoming more and more popular. Designers were playing around with many different designs which made for some strange looking cars. The transition to mid-engined cars and improvements in aerodynamics and downforce meant the cars were faster and the racing was better.

Throughout this era many of the greats battled against one another whether on the same team or not, and the driving lineups were stacked with talent. Feuds and drama were commonplace on and off the track and made F1 even more of a spectacle to watch.

7

Sir Jackie Stewart

Another flying Scotsman in the list of greatest racers. Jackie was born in Dumbartonshire, just outside Glasgow in 1939. His father owned a garage company so he was around cars from a young age. Jackie was not an A* student and left at the age of 15, he was then later diagnosed with severe dyslexia. His talent for racing was not scuppered by this however and once he began racing sports cars he was quickly snapped up by Ken Tyrrell. Jackie began racing in F3 for Tyrrell and won seven races in a row in his first season in 1963. Jackie moved to BRM in 1965 to get a taste of F1 and won two Grand Prix, but when Ken Tyrrell decided to race in F1 in 1968 Jackie left BRM to reform the dynamic duo. In the six seasons Jackie was with Tyrrell he won 27 races and three championships becoming the most decorated F1 driver since Juan Manuel Fangio.

Aside from his race winning pace and championship pedigree Jackie influenced the sport of motor racing in ways no driver ever will. In terms of safety Jackie was by no means satisfied with the regulations and practices in place during his era of motor racing.

With F1 still regarded as the most dangerous sport in the world he used his influence to bring about changes that would save countless lives. Jackie introduced full face crash helmets and seatbelts for the drivers, along with marshals spread around the track and larger ambulance teams to assist drivers if they were to crash. Larger runoff areas on heavy braking zones and safety barriers were introduced to protect the fans. Yet Jackie's push for change did not go without controversy, with many people criticising his campaigns saying he was taking the joy away from the sport. Jackie once stated "It was said that I took the romance out of the sport, that the safety measures removed the swashbuckling spectacle that had been".

Jackie's performances on track spoke for him though. One example is the Nurburgring in Germany, a notoriously dangerous track, where Jackie won four times. Once in 1968 by a whole four minutes in thick fog and rain creating a dangerous surface with poor grip and visibility. These performances gave him a platform to speak from and credibility to discuss the issues of safety.

Jackie's success and personality made him an international superstar, frequently rubbing elbows with musicians, movie stars, politicians and royalty. His glamorous lifestyle never got to his head and he was still a family man raising his two sons with his wife. Making his millions before and after retiring from F1, Jackie was commonplace in corporate boardrooms of huge firms and appeared in many advertisements or promotions. Especially when F1 started to become the worldwide phenomenon that it is today with the arrival of big sponsors as the races were being televised and interest was growing. Jackie was a big part of this change to F1 too, attracting big companies to invest and worked hard to promote F1 on television and spread the word across the globe. Jackie would later receive a knighthood in 2001 for his contributions to motor racing.

8

Niki Lauda

Niki was born In 1949 to a wealthy family who owned several businesses and banks throughout Austria. Niki did not care for his fathers business and when he sidelined university for racing school his father revoked his funding and almost ridiculed his estranged son. This did not deter Niki and paid for the racing school with money he loaned from banks. Buying his way through the ranks of racing thanks to his life insurance policy he secured a seat with March who were unfortunately not competitive enough to allow Niki to showcase his talent. Thanks to his father's dismissal Niki had no other choice but to find a way to land a seat in a competitive car.

In 1973 Niki managed to cook up a deal with BRM which got him a seat and his good performances earned him a new contract that would see his bank loan debt paid off if he stayed with the British company for another two years. Shockingly though, Niki bought himself out of his contract in 1974 using his new employer's money, namely Enzo Ferrari. Ferrari were in a rut and in desperate need for a championship winner as they had been without since 1964. Ferrari admired Niki's no nonsense attitude and his ability to transform a car. Famously on Niki's first test run in his new ferrari he slammed the car saying it was terrible and impossible to drive which given Ferrari's reputation was almost blasphemy, but Niki did not care and vowed to make it a winning car.

This is what he did, in 1975 Niki won his first championship, dominating the season winning in Monaco, Belgium, Sweden, France and the USA. Ferrari were overjoyed at winning a title for the first time in almost ten years. But Niki was not bothered, calling the trophies useless and that they cluttered his home. He disliked the trophies so much that he used to trade them for free car washes down at his local car wash.

In 1976 Niki seemed as though he would walk to back to back championships in his Ferrari however, at the German grand prix Niki's car crashed and burst into flames. Niki was sat in the burning wreck for almost a minute with temperatures reaching over 800 degrees Celsius causing horrific burns to his face and inside his lungs from inhaling toxic fumes from the wreckage. He was rushed to hospital and after a few days was given up for dead and a priest visited to administer the last rites. Lauda came round and miraculously made a recovery through brutal rehabilitation, astonishing doctors who said "he recovered through sheer force of will". Left with huge scars from the skin grafts, Lauda can often be seen wearing a hat to mask the scars as they made many uncomfortable.

That same season Niki raced, only six weeks after his crash, in Italy. Heavily bandaged up underneath his helmet he fought through the pain to finish fourth. Jackie Stewart said it was the most courageous comeback in the history of sport. That season ended in a duel between his rival James Hunt in a Mclaren at the Japanese grand prix but Niki withdrew saying it was too dangerous, giving the title to Hunt. This caused outrage in the Ferrari team and they planned to replace Niki which angered him. He won the title the next year in 1977 with two races remaining hungry for revenge and left Ferrari before the last race of the season being dubbed a traitor.

The following seasons racing in less competitive cars demotivated Niki and he retired in 1979 saying he was "bored of racing in circles". He then started his own airline 'Lauda Air' in which he flew for as one of the pilots. But requiring more money to progress the company Niki returned to racing to fund this new enterprise. In 1982 he returned to Mclaren for five million which was the largest contract ever in the sport. In 1984 he won his third championship by half a point over his teammate and young prospect Alain Prost.

Niki won his last race in 1985 and retired afterwards but never left the sport building up quite the CV as a Ferrari advisor, Jaguar team principle, television presenter, non-executive chairman, board member and board advisor for Mercedes F1 team. He even had a role in persuading the young Lewis Hamilton to drive for Mercedes which would prove to be an excellent decision.

Unfortunately, in later life his injuries from his crash took its toll and caught up with Niki and his health deteriorated. He underwent surgery on various occasions and in 2019 he passed away in hospital aged 70.

9

Alain Prost

Frenchman Alain was born in the Loire region in central France in 1955. A keen sportsman as a child his small stature by no means limited him to entering into a large range of sports such as wrestling and football. Not knowing his direction in life he explored avenues of personal training and perhaps professional soccer before setting his sights on motor racing when on a family holiday he took a liking to kart racing. Leaving school in 1974 he became a full time racer and funded it by tuning engines and distributing karts. In 1975 he won a season in Formula Renault in which he dominated winning two championships and then moved up to F3 which he again dominated and won a further two championships. Now on the F1 radar he joined Mclaren in 1980.

During his first F1 season Alain finished in the points four times but was very accident prone. Worryingly some of these crashes were caused by mechanical failures on the Mclaren which eventually caused Prost to move after one year to Renault. His first win in F1 came at the french grand prix, his home race, in Dijon. Racing in a french car the frenchman handled the pressure and came through. In his first three seasons Prost won nine races but even this could not help the souring relationship between Prost and the management at Renault, who blamed him for not winning a championship in their car. Prost was also unfavoured by the fans who much preferred his teammate and compatriot Rene Arnoux in which Prost had a feud with also. Prost reached boiling point and left Renault and moved back to Mclaren.

During the next six seasons with Mclaren, Prost won 30 races and three world championships, once in 1985 and again in 1986 becoming the first back to back winner since Sir Jack Brabham. Prost unluckily came runner up twice, once in 1988 to his teammate Aryton Senna and again in 1990. This rivalry between Prost and Senna would become one of the most bitter in F1 history bringing out the best and the worst in each driver. Both drivers in the MP4/4 Mclaren, the car that would become the most iconic and dominant cars of that period winning 15 of the 16 races in the 1988 season. The best two drivers in the best car made for a spectacle of racing.

Prost was nicknamed the Professor for his intelligent approach to racing, preferring to save the tires and the brakes in the early stages of the race and then push at the end and take victory. Opposingly, his rival Senna pushed flat out all race long so it took all of Prost's brain racing intelligence to beat Senna. In 1989 Mclaren continued to dominate the field and it was a battle between the two drivers as to who would come first and second. With both Prost and Senna very close going into the Japanese grand prix in Suzuka Senna needed to win in order to keep his championship dreams alive. On the final corner Senna cut to the inside, going for an ever decreasing gap on Prost for first place and Prost turned in closing the door and the two collided. Prost retired but Senna managed to continue and eventually won the race however, was later disqualified for dangerous driving regarding the earlier incident with Prost and handed Alain the title.

Prost left Mclaren after that season and joined Ferrari winning five races and again the title came down to the Japanese grand prix in Suzuka, similar to the previous year. With both drivers in contention for the title it was Prost who this time needed the win to keep his title hopes alive. But this time Senna deliberately crashed into Prost on the opening lap on the first corner, sending them both spinning into the gravel and retiring both cars. Prost was furious and said Senna was "a man without value" and what he did was "disgusting". The year after Ferrari's pace fell and Prost blamed the team for losing their way which inherently got him fired from the team before the end of the season. With no seat for the rest of the season Prost took a break, dipping into commentating before returning with Williams-Renault. Prost would win seven more races and another world championship but with rumour of rival Senna joining the team Prost controversy attempted to stop the signing of Senna to Williams-Renault and he announced his retirement in dispute.

After retirement Prost went back to commentary and punditry while also working as an advisor and test driver for Mclaren before buying his own racing team naming it Prost Grand Prix. But it was a failure and encountered financial and political issues and eventually became an embarrassment to Prost who ended the team in 2001.

10

Ayrton Senna

In Brazil, March 1960, Aryton Senna Da Silva was born. A son to a wealthy family he had no need to provide for his family with money from racing. But when he was just four years old he was given a mini go-kart which sprouted a lust for motor racing. Growing up, young Aryton would look forward to waking up early on a grand prix weekend to watch the racing and soon entered the motor racing scene at 13 years old winning his first karting race. Succeeding this he moved to single seater racing in England where he again dominated winning 5 championships in 3 years. Senna made his debut in F1 in 1984 for Toleman at the Monte Carlo grand prix, a prestigious race in appalling conditions meant his second place finish on his debut proved he was a serious talent.

After his season with Toleman, Aryton decided to move on as he believed his talents required more funding and a more competitive car. In 1985 he moved to Lotus where in 3 seasons he won 6 races. Finding that he had again reached the end of the car's potential he set his sights on Mclaren. In his debut season Mclaren would win 15 of the 16 races and Senna would beat his teammate Alain Prost to the title eight wins to seven. This dominance of the Mclaren MP4/4 meant the rivalry for first place would become very heated between teammates Senna and Prost.

In 1988 the title was close again and came down to the last race in Suzuka where Prost took Senna out on the final chicane to secure his world title. This is when the feud began and turned bitter quickly. The following year in 1989, Senna got his revenge at Suzuka where he crashed into Prost at the start of the race, retiring Prost's Ferrari and Senna's Mclaren and securing the title for Senna. This led many drivers to question Senna's safety to drive alongside with Prost saying Senna "cared more about winning than living". His 'on the limit' driving style was a force to be reckoned with and his seemingly removed fear of death meant he was prepared to push where no driver had pushed before. This created its fair share of close calls and 'elbows out' overtakes, but it made for a spectacle to watch as he would storm through the pack and set blistering lap times.

However, his racing persona was left on the track and off it he was a kind and caring individual and gave much of his fortune from racing to charities. It is thought that the sum sits at around $400 million used to help provide futures for underprivileged families back in Brazil. Senna always lived a full life and explained that he never wanted to live a lesser life succumbing to illness or injury. He said that if he were to have an accident that costs his life he would want it to "happen in an instant"

Tragically, this is exactly what happened in 1994 in Imola where his new Williams flew off the Tamburello corner into the concrete wall live on television in front of millions of fans. The world mourned and his state funeral in Sao Paulo was attended by many drivers and pillars of the F1 community. Namely Alain Prost who was among the drivers to escort the coffin. Senna would become a Martyr of the sport capturing the hearts of many a fan and racer.

11

Michael Schumacher

Michael came from a poor background, his father a bricklayer and manager of a karting track and mother the chef in the canteen at the track. Michael was shown the seat of a racing kart from an early age. With little in the way of financial backing from his parents, Michael's performances on the karting track earnt him backing from affluent onlookers to kick start his career. Winning was what Michael did best and he won his first karting championship at the age of six, going on to win the German and European kart championships. By his 18th birthday Michael had left school to work as a mechanic that was soon sidelined so that racing could become his main priority. In 1990 Mercedes snapped him up after his F3 win as a test driver for sportscars.

Michael made his F1 debut really by chance. Brand new to F1, Jordan Grand Prix was short a driver as Bertrand Gachot was imprisoned after assaulting a taxi driver. Schumacher was given the opportunity of a lifetime and grasped it with both hands qualifying the rookie team 7th on the grid at Spa, one of the most prestigious and gruelling tracks on the calendar. This performance was so impressive that Benetton swept Michael up immediately.

Michael won his first race with Benetton in 1992, again at Spa and over the next four seasons he would win 18 races and two world championships. His first title in 1994 revealed a slightly darker side to Michael's racecraft as he famously collided into the Williams of Damon Hill, who was his closest rival in the championship, at the Australian grand prix. Many thought this controversial manoeuvre was on purpose as it secured the championship for Schumacher. It then tainted the title somewhat. But the following year in 1995 Schumacher won the title without question.

Ferrari were a team with no direction and were calling out for a knight in shining armour. When Schumacher arrived in 1996 championships did not arrive straight away. Winning three races in 1996 and five in 1997 it looked promising but another ramming manoeuvre on another Williams, this time Jacques Villeneuve, cost Michael the title this time through punishment from the FIA who struck him from the standings. Second in the championship in 1998 and interrupted by a broken leg in 1999 saw Michael come close again. Many were wondering if Schumacher was capable of winning it all in the Ferrari and in 2000 the floodgates opened.

In 2000 Schumacher won his first title with Ferrari, their first in 21 years. Michael evolved into the most dominant driver in the history of the sport winning the title in 2001, finishing on the podium in all of the 17 races in 2002 and winning his 6th title in 2003, breaking Fangio's record that stood for almost 50 years. 2004 dominance again with Michael winning 13 of the 18 races and winning his 7th title by 34 points.

Michael was naturally blessed with an ability to think quickly and handle multiple sources of information at one time. His huge efforts off of the track with his conditioning meant he was the best trained athlete on the grid and meant he could sustain his intensity throughout the race. He knew where the limits of his car were and rarely made a mistake.

Alongside being an incredible athlete Michael was a caring man, often visiting the Ferrari factory to thank the workers and boost morale, he said he loved the Ferrari family. He also gave generously to charity and in 2004 Schumacher made a personal donation of $10 million to the Asian tsunami cause.

Schumacher retired in 2006 after coming second in the championship winning 7 more races and bringing his total to 91, which was 40 more than second place Alain Prost. Unsurprisingly, Schumacher could not keep himself out of the driving seat and returned in 2010 aged 41 with newly formed team Mercedes headed by good friend Ross Brawn. Some say it was a mistake returning and tainted his career only finishing on the podium once during his three year return. Schumacher retired for good in 2012 concluding that "now is a good time to go" as he was surrounded by championship winning drivers that were 10 years younger than himself. But saying that it was a good learning experience, commenting that "losing can be both more difficult and more instructive than winning".

Having survived so many seasons in one of the most dangerous sports in the world it came as a shock to the world when, in his first year of retirement, Schumacher suffered a very serious head injury while on a family skiing holiday. This injury left him in a coma for many months but when he awoke he was never the same again. A cruel twist of fate to one of the greatest sportsmen of all time.

12

The Modern Era

Now that F1 is cemented as a worldwide phenomenon, televised across the globe with fans in all corners of the world. Billions are invested into state of the art equipment and facilities to create the most technologically advanced racing cars ever seen. The turbo hybrid era is the most recent transition seen in the sport with cars now drawing power from an electrical component as well as a standard combustion engine. With a focus on their contribution to climate change, more effort has gone into reducing the carbon footprint of the sport, developing more eco-friendly fuels and improved efficiency. The 2021 season saw the greatest number of watching fans in the history of the sport, with netflix documentaries like "Drive to Survive" providing a greater insight into one of the most exciting sports. This has spread the word to a wider audience making for the subsequent seasons to be the most exciting yet.

13

Sir Lewis Hamilton

Born Lewis Carl Davidson Hamilton to a mixed race family in 1985, Stevenage, United Kingdom. Lewis lived with his mother Carmen and his father Anthony who's parents immigrated from Grenada in the 1950s. When Lewis was just two his parents divorced and Lewis went off with his mother for 8 years until he moved in with his father Anthony. Living with Anthony's wife Linda and son Nicholas, who had cerebral palsy, Lewis has frequently commented on how he draws inspiration from his step brother. Lewis says he draws a caring nature from his step-mum and his strength and determination from his dad.

Coming from a poor background, Lewis' dad Anthony worked tirelessly to promote and further Lewis' racing career, working three jobs to get the funds required. When Lewis was eight he bought a well used go-kart that cost one months income. But this was the beginning of Lewis' very successful career.

In 1995, Lewis was a karting champion at 10 years old. At the event was Mclaren Mercedes F1 boss Ron Dennis who was approached by the young Lewis with an autograph book. Lewis asked for a signature and said "Hello Mr Dennis, I am Lewis Hamilton and one day I'd like to race for your team". Ron obliged and added his number and said to "call him in 9 years".

However, it was only three years later that Ron called offering to sponsor the young Lewis Hamilton in his racing endeavours to take the toll off of his parents, as long as Lewis stayed focussed in school.

The removal of the financial burden was a blessing but it did add extra pressure and scrutiny onto Lewis' career, with expectations even greater now that such a prestigious team had taken such interest so early in his career. But Lewis rose to the occasion and was a worthy investment winning 8 championships in 6 years of kart racing and 3 major single seater titles. Many of these wins backed with performances from the back of the grid to the front. No wonder Lewis was soon promoted to the Mclaren F1 team in 2007.

This rookie season will go down as one of the greatest, consistently out-performing teammate and previous world champion Fernando Alonso, winning four races and led the championship for 5 months only to lose the title to Kimi Raikkonen on the last race by a single point. This debut season set him up for stardom however the next few years were not good for Mclaren with Lewis and Fernando becoming bitter rivals and the Mclaren team being investigated for trading secrets with Ferrari meant that their drivers championship points were struck from the record and team fined $100 million

2008 was a better season and with Alonso out of the picture, Lewis was team leader and won 5 races and 6 podiums which took the world championship to the final race in Brazil. Hamilton had to finish 5th or better or the title would go to Brazilian Felipe Massa at his home race. The race was thwarted by rain and Massa handled the weather perfectly finishing 1st. Hamilton was 6th on the last lap and in a thrilling battle to the finish Lewis managed to overtake on the last corner to come home in 5th and win his first world championship

Lewis then rocketed to celebrity status, dabbling in pop culture, fashion and music. Lewis made himself a name off the race track starting fashion labels and mingling with the rich and famous. Lewis was also the first black driver and world champion in the history of the sport and he used his status to promote equality, being a big advocate for black lives matter and is an avid vegan, setting up his own vegan restaurant in London.

Following his first championship winning season Lewis struggled to maintain that world beating form and fell off the pace crediting messy breakups and a stressful private life for distracting him from racing. In 2012 he regained his pace with four wins but announced he was leaving Mclaren before the season was over. He joined Mercedes in 2013 replacing the retiring Michael Schumacher and in his first season he won one race and came fourth in the world standings. But it was the following season things really started to click for Lewis and Mercedes. With new regulation changes coming to the sport, the cars changed dramatically with new setups and engine types meant less downforce and new hybrid engines

Mercedes nailed the new regulations and their new car was a cut above the rest for several seasons. Lewis and his teammate Nico Rosberg were in their own races and with no team orders they were allowed to race each other for the race win week in week out making for some dramatic battles and some heated post race exchanges. Lewis and Nico were friends from karting and their bond was put to the test during these years as teammates. In their first season together in these new cars in 2014 went down to the final race but Lewis came out on top in Abu Dhabi. 2015 was more of the same in the way of Mercedes dominance, but more in favour of Lewis winning 16 of the 19 races. As Lewis had more failures and driver errors than his counterpart, 2016 was Rosberg's turn winning the title by five points. Rosberg retired after his win leaving Lewis to dominate

Unstoppable from 2017 through to 2020, Lewis and the Mercedes were completely untouchable, with new teammate Valtteri Bottas struggling to keep up in an identical car. Lewis collected his fourth five sixth

and seventh title in from 2017 to 2020 equalling the great Michael Schumacher becoming the tied greatest racing driver of all time. With his sights on his eighth

It was only until the 2021 season that Max Verstappen could come close in the RedBull to Lewis and in one of the greatest seasons the sport has ever seen. Lewis had his 8th title snatched from him on the last lap of the last race in Abu Dhabi in controversial fashion when a safety car was called out with a few laps to go. Not wanting to sacrifice track position Lewis stayed out on old tires but Max pitted for fresh rubber, in a controversial decision from race control only the cars between Lewis and Max were allowed to unlap themselves meaning Lewis was at a huge disadvantage thanks to his old tires. Max overtook on the race restart and won the title

But this season revealed Lewis' character, with his composure when the result does not go your way proved what a sportsman Lewis really is. Tied for the most titles in the history of the sport with years worth of racing left in him it would not be surprising if Lewis went on to be the all time most decorated racer

14

Sebastian Vettel

orn in July 1987, in Heppenheim south west Germany. Sebastian Vettel's dad was a karting enthusiast and occasionally competed in hill climb races, so it was not a surprise when young 'Seb' was introduced to a kart at an early age. Making his karting debut aged seven, Seb impressed with many wins and championships, even a certain Michael Schumacher who presented young Seb with one of his trophies. Schumacher was Seb's idol in motorsport and the two became close due to their similar backgrounds, Schumacher encouraged Seb to aim for the top.

Vettel's big time shot came when Red Bull talent scouts decided Seb was fit for their young driver talent program. At 17 years old Seb won 18 of his 20 races in single seater racing in the German Formula BMW championship. Sticking with BMW Vettel test drove for BMW Williams F1 team at 18 years old, and at 19 was testing for BMW Sauber F1 team. His first taste of F1 was in 2007 at the US grand prix in Indianapolis replacing the injured Robert Kubica. Vettel qualified 7th and finished 8th becoming the youngest ever driver to earn a championship point. Mid-season he signed with Toro Rosso, Red Bull's second team. In 2008, during his first full season with Toro Rosso, Vettel earned his first F1 pole and win in Monza to become the youngest ever race winner and pole sitter.

In 2009 he moved up to Red Bull alongside the experienced Mark Webber who was 10 years his senior, but age did not help Mark outpace his young teammate who won 4 races and came 2nd in the championship behind only Jenson Button. 2010 was the longest season in the sport but Vettel showed focus and resilience and brought home the title. Even against close rivals Ferrari and Mclaren who's drivers Alonso, Hamilton and Button all led the championship at some point. But it was young Seb who's race pace trumped the rest making him the youngest to win a title at 23 years and 133 days.

In the following seasons Red Bull came to dominate, with their car being a cut above the rest in 2011 making Vettel's title defence a walk in the park winning the championship with 4 races remaining, making him the youngest ever back to back champion. 2012 was much closer with five other world champions fighting for the title but again Vettel came through and beat Alonso by 3 points to become the youngest ever triple champion. In Brazil coming back from last place to complete a comeback of four wins in a row plus three podiums to clinch the title on the final race. In 2013 Vettel won his fourth championship winning 10 races with 6 in a row. By the end of the season he had won 9 races in a row equalling a 60 year old record.

By 26 years and 123 days old Vettel had won 4 titles in a row and was expected to carry on and rewrite the record books, sadly that was not the case. In 2014 Vettel failed to win a single race and subsequently left Red Bull to join Ferrari which picked up well initially. Vettel was best of the rest in an era dominated by Mercedes and rejuvenated a slacking Ferrari team in 2015. But this joy did not last long and the results failed to come in during subsequent seasons with mistakes and inconsistencies creeping in costing him points meant he finished fourth in the standings. 2017 was a revival with Vettel where he led the standings for the first half of the season and in the end only coming second to Lewis Hamilton in the Mercedes. In 2018 Ferrari had a good car but Vettel could not get the consistency to win the championship, with costly accidents occurring again. Desperation began to creep into Vettel's driving and unnecessary contact and crashes made for unreliable performances and a volatile Seb.

Much to his dismay, his new young and up and coming teammate Charles Leclerc took the car quickly in 2019 and outperformed Vettel. This did not sit well with Vettel and his relationship with Ferrari began to turn sour with strategy mistakes and errors on his part meant that in 2020, halfway through the season, Ferrari announced that Vettel's contract would not be renewed for 2021

Taking a dent in his pride Vettel left Ferrari after their worst season in their history for new beginnings at newly formed Aston Martin. Vettel was hopeful joining this new team as a "new adventure" and believes the "future looks even brighter"

15

Authors Notes

I sincerely hope that you enjoyed reading this book. I imagine that it took you back and reminded you of some names that you have not heard for a while, cemented your opinions, or even introduced you to some new names. Writing this book was a load of fun and I hope you could sense some of that in the book. It excites me to see how far this sport has come and for what the future holds.

It would be greatly appreciated if you could leave a kind review on this book, that would help me no end.

See you in the next one.